Water is all around us.
Water is in ponds, rivers, lakes,
and streams. It is also under
the ground! People, animals,
and plants all need water.

Why do we need to have
fresh water?

We need water to drink. A tall glass of water is always so good! We also need water for washing. We use it to wash our pets!

How do you use water?

Some people swim, which helps keep them strong and healthy. This father is teaching the small boy to swim.

When have you had fun with water?

Water helps keep us safe.
When a fire starts, people call
the fire squad. The squad uses
water, which helps stop the fire.

If you need the fire squad,
call 9-1-1!

A lot of the earth is water. Sometimes water falls down over rocks. This is called a waterfall.

Could you make a small waterfall?

Sometimes water falls from the sky. White clouds turn dark. Then rain falls to wash the earth and water the plants.

Why do plants need water?

Sometimes water vapor freezes and forms snow. Snow falls, coating the earth with white.

Why might you like to go sledding?

Water is all around us. We all need water, which helps us in many ways.